An Introduction to Einstein's Relativity
Version A: Math Omitted

W. Blaine Dowler

April 3, 2025

Cover design by: Art Painter
Library of Congress Control Number: 2018675309
Printed in the United States of America

About This Book

The contents of this book were originally published in a series of nine weekly lessons at http://www.bureau42.com in the summer of 2012. Those lessons were released in two versions, as is this book. One version includes all of the math. The other, which you are holding, omits the math. Furthermore, this version cleans up some explanations and corrects some typos from the original web-based version.

Contents

Chapter 1

The Need For Relativity

1.1 The Success of 19th Century Physics

By the end of the 19th Century, physicists were standing proud. It seemed as though every unanswered question in physics was on the verge of being answered, and that physics would be the first field of science to be described in totality. The list of unanswered questions was short, and the answers seemed to be a matter of applying basic mathematics to several ideas and compiling the results. We shall now explore some of those specific questions in detail.

1.2 Mercury's Orbit

Most of us imagine planetary orbits as circles, or even ellipses, which wrap around a star. The planet then travels along this fixed path, over and over, repeating its exact positions again and again.

Most of us imagine this incorrectly. As it turns out, these orbits precess, meaning the orbit itself rotates around a star. In other words, if you were to take a piece of paper and plot the orbit over time, instead of going over the same line over and over again, it would shift ever so slightly, and produce more of a "floral" pattern as the planet's orbit itself revolves around the Sun. Newtonian mechanics predict such a shift. In the case of Mercury, however, the shifting and precession predicted by theory were not enough to explain the degree of shift found by observation. The orbit spins more rapidly than anyone had predicted.

1.3 Speed of Earth and the Ether

In the physics of the 19th century, it was thought that light was a wave with no particle properties at all, and that it couldn't travel without some sort of material or medium to travel through. As it was clear that light reached Earth from the Sun, and as the finite extent of the atmosphere had been established, that meant there was something out there in space that wasn't air between the Earth and the Sun. Thus was born an idea in physics that made so much sense at the time that some people still won't let it go today: the ether.

The ether was a material thought to be of exceptionally low density that spread throughout the heavens, and it was this ether that light travelled through. As it seemed to have no measurable impact on orbital mechanics, it was also suggested that all of the ether moved or flowed at a steady speed, and that planets sailed through it like a gentle ocean current. This begged an obvious question: how fast is the current?

The speed of the current could be measured with a relatively simple apparatus. Light waves have peaks and valleys, technically called crests and troughs, just as water waves do. As with water, when crest meets crest, they combine to form a higher crest. Similarly, two troughs combine to form a deeper trough. This is called constructive interference. When two waves are "in phase," it means their crests and troughs always combine in this manner. When the crest of one wave combines with the trough from another, they cancel each other out through a process known as destructive interference, and are said to be "out of phase."

This principle was applied to measure the speed of the Earth relative to the ether. Imagine two swimmers in a lake, starting at a floating platform. Now imagine that they are in a race, and have to swim from the floating platform to a landmark in the lake and then back to the platform. Further imagine that each swimmer has his or her own landmark to swim to, and that the landmarks are the same distance from the platform, but that they are in different directions, separated by a right angle. If the swimmers swim at the same speed, they would be expected to swim to the landmarks and back in the same amount of time. If the swimmers were waves, they'd arrive "in phase."

Now take this same setup with our swimmers and put them in a river. Even travelling at the same speed, if one swimmer swims with and against the current, and the other swims at a right angle to the current, they would not be expected to return to the floating platform (which also drifts with the current) at the same time. This is because the flow of the current would ensure that the swimmers were moving at different speeds. If the swimmers were waves, they'd arrive "out of phase."

Albert Michelson and Edward Morley decided to use this principle to measure the speed of Earth relative to the ether. They set up an apparatus using one spinning mirror and two stationary mirrors to split a single beam of light into two beams that travelled at right angles to each other before reforming. They could then measure how "out of phase" the light beams were upon their return and recombination, and determine the speed of the apparatus with respect to the ether along a particular direction at that time. By repeating the experiment several times in the course of a year, the Earth's motion relative to the Sun could be measured and eliminated, and the motion of our solar system relative to the ether could be quantified.

The problem they had was this; no matter what time of year they did the experiment or what size apparatus they used, the light returned in phase every time without exception. For some reason, the speed of light wasn't changing, with or without an ether. It appeared that the entire theory of the ether was fundamentally flawed and that there was no ether. That's not the conclusion most scientists reached, however. The majority of the thinking was not in looking for an alternative to ether, but rather in looking for a flaw in the Michelson- Morley experimental design, despite the number of independent confirmations that were coming in from those who reproduced the experiment. Others looked at creating a more complicated picture of the ether which would have the same result.

1.3.1 The Ether Today

Every reproducible experiment conducted over the past 125 years indicates that the ether does not exist. Despite this complete lack of supporting evidence, there are still those who cling to the theory. Some seem to do so for the novelty, as it has almost become a game to discern how complicated ether must be in order to remain consistent with all of these experiments. Others do so for other reasons. For example, ether is a fundamental piece of most theories which attempt to include existence of ghosts and other paranormal activities in modern science.

In short, people still talk about theories involving ether despite a complete lack of supporting evidence. If you encounter a theory involving ether, take it not with a grain of salt, but with an industrial size salt lick.

1.4 Compatibility of Newton and Maxwell

In the late nineteenth century, the world of physics' most legendary figure was Isaac Newton. Newton's laws of motion had been formulated in the 17th century, and stood unchallenged over 200 years later, having ushered in a new era in predictive, quantitative science. They agreed with experiments to an amazing degree.

In the latter half of the 19th century, the theories of electricity and magnetism started to mature and formal structures started to appear. Through a combination of deriving some relationships and compiling the work of others, James Clerk Maxwell assembled the so-called "Maxwell's Equations," which described electricity and magnetism effectively.

Now, Maxwell's equations had been amazingly successful. Not only did they accurately describe the behaviours of anything involving both electric charges and magnetic fields, they also allowed Maxwell to make the first (reliable) prediction of the speed of light. Light is a combination of electric and magnetic fields that feed off of and propagate each other. Using only some physical constants related to these waves, Maxwell calculated the speed of light.[1] Before that time, the speed of light was a quantity known only by experimental measurements, and had a considerable margin for error. Maxwell's calculation came very close to the now accepted value for the speed of light. Not only was it the first theoretical value for the speed of light, it was a value high enough to explain why the experimentalists had a hard

---

[1]To be pedantic, as any good mathematician or theoretical physicist must be, Maxwell actually calculated the speeds of electric and magnetic fields independently and then proved they were not only identical, but that the two calculations were closely related. It was this consistency in both result and procedure that helped cement his findings in the minds of his peers.

time computing the value correctly. In these early days, the work of Newton and Maxwell appeared to join together like a marriage made in heaven.

In spite of these triumphs, there was a serious problem. To illustrate this problem, we must first discuss reference frames.

1.4.1   Reference Frames

In science, all experiments require a well defined reference frame.[2] It is, basically, a set of standard positions and directions that are used to measure things against.

For example, imagine you are doing the first physics experiment in a high school physics course. This usually involves dropping something from a given height, and measuring how much time it takes to fall.[3] The distance fallen is typically measured relative to the ground, and the analysis of the accelerations and speeds involved are all relative to the laboratory bench and floor.

This typical and effective setup completely ignores the fact that the Earth rotates during the day about its axis, and the entire planet is revolving around the Sun. The entire solar system also orbits around the centre of the galaxy, and our galaxy may be in motion with respect to some other as yet undetermined body. These motions can be safely ignored in the high school laboratory.[4] We have chosen that laboratory as our frame of reference; we measure with respect to a point in the lab, and can continue unrestrained from that perspective. This is a valid assumption, provided that all frames of reference are equivalent from a scientific standpoint.

Note that we use the word "equivalent" instead of "equal" here, and with good reason. If two reference frames are equal, the quantities measured in one will be an exact match to quantities measured in the other. If they are equivalent, then observers in the two frames would be able to accurately predict future events, even though they measure different numbers for the quantities involved. For example, an observer in the high school laboratory might say that an object has zero speed, velocity, and energy of motion before it is dropped, while an observer on the Moon (which is moving relative to the Earth's surface) would disagree. Both observers would be able to accurately predict what would happen when the object is dropped, so the frames are equivalent, but not equal. Specifically, they both predict that the falling object would take the most direct path possible when moving towards a collision with the laboratory floor.

In Newton's world, these frames are equivalent. When Maxwell's equations are combined with Newton's theories, we find that two different frames of reference are

---

[2]The term "reference frame" is interchangeable with "frame of reference." The latter is somewhat old fashioned, and is not seen as often these days.

[3]Physics experiments at the high school level can be painfully boring. I can assure you, they do get better. Eventually. We need to start students with the most basic experiments to make sure they have their experimentation skills down well enough to move on to the exciting, interesting experiments, which may or may not involve violent explosions and supersonic projectiles.

[4]The Earth's rotation does have an impact measurable by sufficiently advanced equipment, but we'll leave that discussion for later. If it bothers you now to ignore this piece, rest assured that the Earth's motion could be completely ignored if it were linear instead of rotating, and that high school equipment budgets make it extremely difficult to design high school experiments capable of detecting the difference.

not equivalent.

1.4.2 How Charged Objects Interact

Experiments had confirmed a number of behaviours of electrically charged objects. Maxwell's equations formalized these behaviours, but that's all they did; interactions between magnetic fields and electrically charged objects were jammed into the theory because they were observed in the lab, not because they were predicted by existing theory.

When charged objects are stationary, the basic phenomena are explained in a very simple fashion: unlike charges attract, while like charges repel. If you arrange a number of electrically charged marbles on an insulated surface and prevent them from moving, Maxwell's equations will accurately predict the attractive and repulsive forces they all experience.

When charges move, things get complicated. Moving charges produce magnetic fields[5] which can also attract or repel. These behaviours depend on the directions the charged objects are moving as well the charges themselves. Specifically:

- Like charges moving in the same direction produce attractive magnetic fields.

- Like charges moving in opposite directions produce repulsive magnetic fields.

- Unlike charges moving in opposite directions produce attractive magnetic fields.

- Unlike charges moving in the same direction produce repulsive magnetic fields.

The faster the charges move, the stronger the magnetic fields they produce, and the more pronounced the forces become. The same is not true of the electric forces: they are entirely independent of the relative motions of the particles.

1.4.3 Newton to Maxwell: "It's Not Me, It's You"

Imagine two wires, running parallel to each other for as far as the eye can see. In fact, imagine them to be running farther than the eye can see; imagine they are infinitely long. Now, imagine that they carry identical electrical charges, and that these charges are spread uniformly across the wires. No one point on either wire is distinguishable from any other point on the same wire.

An observer standing next to the stationary wires would predict a repulsive force, and the two wires would move apart. That is the result of the laws of physics as measured in his or her reference frame.

Now imagine a second observer flying along parallel to the wires, exactly half way between them. From this observer's reference frame,[6] the wires are in motion, and

---

[5]Why do moving charges produce magnetic fields? In Maxwell's time: "because they do." This was discovered in the lab, and subsequently jammed into the theory because it needed to be there, and not because any theorist could predict or explain the phenomenon.

[6]An observer's reference frame is one centred on that observer in which the observer is always treated as stationary.

moving in the same direction, so the wires experience *two* forces. In addition to the repulsive electrical force, there is also an *attractive* magnetic force. The electric force is independent of the relative motions of the wires. The magnetic force, however, increases with speed. As our flying friend accelerates, the attractive magnetic force gets stronger and stronger, until it eventually overpowers the repulsive force and the wires approach each other instead.

This is a significant discrepancy. Do the wires move apart, as predicted by our standing observer, or do they collide, as predicted by our flying observer? These predictions are *not* equivalent, and so neither are the two frames of reference.

Newton's theories were those that not only governed motion, but which also described how to work in moving frames of reference. His theories had stood up to every conceivable experimental test for more than 250 years. As a physicist, he was revered by those working in the fields of research, and seemed more than human through the lens of history.[7] Maxwell, on the other hand, was very human, and his equations were relatively recent discoveries, having been collected and studied for less than 20 years. Thus, the overwhelming majority of physicists turned to Maxwell, looking for his mistake, and thereby fixing the problem with incompatibility.

Now, physicists are human, and when a theory that stands as long as Newton's is shown to be incompatible with something new and exciting, there are two natural reactions. The first, and by far most common, is to assume that the older Newtonian theories are correct and the new ones are wrong. However, in the world of faceless science, there are young upstarts excited about the prospect of becoming famous by rewriting the "old and busted" theory to force it to conform to the "new hotness" theory. One such upstart was Hendrik Antoon Lorentz, who, at the tender age of 51, decided to determine exactly how to rewrite Newton to make his work compatible with Maxwell's. His work resulted in an absolute absurdity: to reconcile the two theories, time had to depend on the speed of the observer! To most, this constituted further proof that the problem lay in Maxwell's work.

Unfortunately, no problem could be found. Many scientists started connecting the problem to theology, deciding that somewhere out there was a reference frame that worked. In other words, they believed that detailed explorations and experimentations would reveal situations that were not consistent phenomenologically with a laboratory based experimental reference frame. By determining which reference frame was consistent with such experiments, they would be able to determine God's preferred frame of reference. They intended to then search the heavens for such a body, in hopes of discovering the location of Heaven. The early results of this search were inconsistent and fruitless.

1.5 Enter Einstein

The world of professional physicists scoured the experimental and theoretical worlds looking for solutions to these problems. It took a self-taught physicist employed as

---

[7]The lens of history is about the only lens Newton didn't completely describe up to that point, which was likely a contributing factor.

a patent clerk to see something nobody else had seen in the work. Most physicists were excited because "Maxwell had calculated the speed of light using only some *physical constants*!" It was Einstein who got excited that "Maxwell had calculated the speed of light using *only* some physical constants!" Einstein realized that certain expected information was absent from the calculation. The full implications of this realization and the birth of the special theory of relativity are the subject of the next chapter.

Chapter 2

The Revelations of Einstein

2.1   Einstein's Revelation

In the early 1900s, physicists were confused. Electricity and magnetism, as described by Maxwell's equations, seemed incompatible with Newton's work in motion. The incompatibility arose when two different observers studied the same phenomenon from two different frames of reference. On top of that, they had a hard time studying the ether they had proposed as a fundamental part of their theory of light, as they could not produce the interference effects one would expect when the speed of light changed with respect to the medium the light was travelling through.

In 1905, Einstein solved both of these problems with one revelation.[1] Scientists were thrilled with Maxwell's calculation of the speed of light, which was derived solely from the values of a few physical constants. Einstein was the first to recognize that a fundamental piece of information appeared to be missing from the calculation. Maxwell's calculation of the speed of light was completely unrelated to the observer's frame of reference.

In other words, two different observers looking at the same light wave would agree on the speed the light was moving, regardless of how they were moving with respect to each other. They do not necessarily agree on the velocity, as that includes direction, but they will agree on the speed.

The manner in which this explains the Michelson-Morley result is relatively clear: if you cannot produce a difference in the speed of light along the two paths, you cannot produce different interference patterns by moving the apparatus. The manner in which this reconciles with Maxwell is far more involved; we will need to lay some groundwork before we can cover that.

---

[1]The third problem mentioned in the previous lesson, dealing with Mercury's orbit, would also be solved by Einstein, but not for another ten years.

2.2   Postulates

In mathematics, when one is formulating a new theory or idea, one begins with "axioms," or ideas that are assumed to be true. The equivalent statements in the experimental sciences are called "postulates." A postulate is something that seems reasonable and is assumed to be true based on daily experiences or scientific observations. A properly formed physics theory begins with postulates that are thoughts formed in some spoken language, which is then translated into mathematics and the implications are followed.

2.2.1   Newton's Postulates

When Newton did his work in motion, he proposed several postulates, including three laws and six corollaries. He also made a few assumptions that he did not formally propose as a part of his system. The laws and his key assumptions are listed here:

1. Law 1: Unless an outside force is applied, a stationary object (object "at rest") will remain stationary and a moving object will continue moving at the same speed and in the same direction.[2]

2. Law 2: If one does apply a force to an object, that object will accelerate in the direction the force is applied. Greater forces create greater changes in speeds. These changes are proportionate to both the force and the "quantity of matter" in an object. Newton originally proposed that the "quantity of matter" was the inertia of an object, and then suggested that the mass of an object is that inertia, so that mass and inertia were identical, which is why inertia is discussed conceptually in class and then mass is used in the algebra.

3. Law 3: If body A pushes against body B with a given force in a given direction, body A also feels a force of equal magnitude and opposite direction coming from body B.

4. Unstated assumption 1: In all of Newton's work, though he didn't formally state it, he assumed that two different individuals measuring the elapsed time between two events would always measure the same interval of time. In other words, if some poor kid's scoop of ice cream falls off its cone at a fair, the time it takes that ice cream scoop to fall to the ground would be the same amount of time whether measured by that child's parent or by someone riding the ferris wheel.

5. Unstated assumption 2: This, technically, was unstated in the Principia Mathematica in which he detailed his laws of motion, but it was written in Newton's

---

[2]Prior to Newton, it was believed amongst the Christian cultures that moving objects required continuous outside forces to stay in motion. Objects moving along the ground stopped not due to friction but due to the lack of a continuous outside driving force. The stars, planets, moons and other heavenly bodies remained in motion because, it was believed, God had tasked angels with the job of pushing them constantly around in the sky for the amusement of humans. The author is unaware of any suggested reason that God disliked these angels so much that He doomed them so such a monotonous eternity, nor is he aware of alternate theories proposed by the non-Christian cultures.

work in theology. Newton recognized the possibility of nonequivalent reference frames, and was one of the first to propose that a "preferred" reference frame could be found that would indicate the location of the Christian Heaven.

The corollaries are clarifications of how the math behind the three laws works, and will not be discussed here in detail.

2.2.2 Einstein's Postulates

Einstein proposed two postulates in 1905 that turned the physics world on its head, eventually. The initial reaction to Einstein's crazy ideas was utter disbelief in response to some of the mathematical implications.

1. All "inertial" reference frames are valid and equivalent. An "inertial" reference frame is one in which the origin of the coordinate system is not being accelerated. Thus, you can measure relative to a stationary object or to an object moving at a constant speed, but not relative to someone riding a ferris wheel, since that person's direction is changing. (Direction counts for acceleration.) In 1915, Einstein would formulate a new theory that would allow the use of any reference frame, treating them all as equivalent, but that addition will come in a later lesson.

2. The speed of light is "invariant," meaning all observers in all valid reference frames measure exactly the same value for the speed light is moving at.

Einstein felt he was on the right track when he used these postulates to derive new ways to adjust the math for changing reference frames, and arrived at a result *identical* to the one that H. A. Lorentz found when he tried to reconcile Maxwell's equations with Newton's laws. That seemed to be a very compelling coincidence.

The complete implications of these postulates are not obvious, particularly if one is trying to see the connection between these ideas and the famous $E = mc^2$ equation associated with relativity. We will examine some of the implications of these postulates from now until the end of chapter six. In chapter seven, we will add the postulates that allow reference frames to experience acceleration, and deal with those implications for the rest of this book. These early chapters deal with the special theory of relativity, since they relate to the "special case" in which gravity and other outside fields and forces do not apply. The latter chapters deal with the general theory of relativity, in which gravity and accelerations are included, so that we can work with "general" cases of physical situations.

2.3 Implications for Distances

Let us begin by examining the way distances are measured by moving observers under Newtonian mechanics and reference frames.

Imagine we have two observers, named Hal and Barry. Hal is standing still, while Barry is running at a constant speed. Both are observing a car driving in the same direction Barry is running and at a higher speed than Barry is running with. When Hal measures the distance the car has travelled, he measures relative to himself.

Barry also measures the distance the car has travelled relative to himself, but he is *between* Hal and the car, so he concludes that the distance travelled by the car is less than the distance Hal measures.

"But wait!" one might cry. "Why not measure relative to the road to eliminate this problem?" This is what is commonly done in labs for consistency. However, the important thing to recognize is that there is *no* problem here. If all inertial frames are equivalent, then it doesn't matter who does the measurement, provided the frame of reference is described accurately in the results. This helps in cases such as the motion of astronomical bodies, in which case there is no clear "road" to use as a point of reference.

On a conceptual level, there is no difference between the Newtonian view and the Einsteinian view. The observer moving in the same direction as the car will measure a smaller distance travelled than the observer who is stationary relative to the road. Once the math is examined in detail, one finds that the difference between the two measurements is more dramatic in the Einsteinian view, but the basic concept is the same. The difference is due to the invariance of the speed of light; there are no "instantaneous" measurements, since information travels no faster than the speed of light. In short, Hal and Barry don't measure the position of the car "now," but instead measure the position of the car according to where it was when it reflected the light their eyes are detecting "now." This lag causes the difference between the Newtonian and Einsteinian measurements observed here.

This has further implications. A car has a volume; what if they measure the distances to both the front and rear of the car? The time lag due to the transit time of light is more pronounced from the front (further) side of the vehicle than it is from the rear (closer) side. Thus, if you ask both Hal and Barry to measure the length of the car, they would provide two different answers! In relativity, objects moving relative to the observer contract or shrink, and do so in greater degree as the differences in speed become more pronounced. This is known as *length contraction*.

2.4   Implications for Time Intervals

Under Newtonian mechanics, Hal and Barry would disagree about the distance the car travelled in our previous example, but they would agree on the amount of time the car spent in motion. The invariance of the speed of light throws a monkey wrench into that perspective.

Imagine the same situation as above. This time, however, the car's headlights are on. Instead of measuring the distance and time elapsed of the car's travel, Hal and Barry measure those quantities for the light waves emitted by the car's headlights.

Just as before, they measure the distance travelled by that light, and come up with different results. In Newtonian mechanics, they would each measure the speed of that light by measuring the elapsed time, and then dividing the distance travelled by the elapsed time. As they would agree on that time interval, they would arrive at two different numbers for the speed of light. In Einstein's relativity, this is impossible; all observers measure light in a vacuum to be travelling at the same, invariant speed. What implications does that have?

We can flip the math around. Both observers measure the distance travelled, though they disagree on the exact value of this distance. They also measure the speed of light, and both arrive at the same value. The time elapsed can be calculated by taking the distance travelled and dividing by the speed of light. Hal and Barry each do this calculation, using *different* numbers for the distance travelled, but the same numbers for the speed of light: they arrive at *different* numbers for the time elapsed, each of which is correct in that observer's reference frame!

This was a staggering thought. Time depends upon the speed of the observer. This phenomenon, known as *time dilation*, has since been verified by many, many experiments. This is a real and undeniable effect. If someone observes a moving clock, he or she would conclude that the clock in motion is running more slowly than an accurate clock which is considered to be "at rest."

2.5   Implications for Velocities

Let us return to Hal, Barry and the car. This time, instead of running in the same direction as the car, Barry runs in the opposite direction. Let us also assume that, as Hal measures things, both Barry and the car are moving at more than half the speed of light; for the sake of argument, let us choose three quarters of the speed of light as the speed they travel.

In the Newtonian view, if Barry were to measure the speed of the car, he would calculate the speed that is simply the sum of his speed and the car's speed, as measured by Hal. In this case, that would result in an answer that is one and a half times the speed of light. In the Einsteinian view, this changes dramatically.

For the sake of argument, let us assume the car's headlights are still on. When Barry measures the speed of the light coming from those headlights, he concludes that it travels at the speed of light, as he must. When he measures the speed of the car itself, he finds that it is slower than the speed of light; the light wave from the headlights are further away from the car, and getting further away all the time.

In other words, the Einsteinian view allows for agreement about *relative* speeds of bodies, but not *absolute* speeds. Simple comparisons such as "faster" and "slower" are maintained, but the actual numbers vary considerably. This is because the observers don't just disagree on the distances travelled, but also on the time elapsed. The truly stunning implication is this: nothing can travel faster than the speed of light. Any moving object can "turn on its headlights" or not, and it will not affect how others measure the speed of that body. Everybody will, however, agree that the hypothetical light from those headlights travels faster than the source (the moving object) and that the light is travelling at the speed of light. Therefore, all observers agree that the source of the light is moving slower than light.

2.6   Einstein and Maxwell

Now that these simple implications have been discussed, we can return to our paradox from electricity and magnetism. Recall the situation with two infinitely long and identically electrically charged rods. From a reference frame in which they are stationary, the rods will repel each other due to the like charges on the rods. The more dense the electrical charges are on those rods (i.e. the more charge is present in any given length of the rod) the stronger this repulsion becomes. From a moving reference frame, these charges form a current, which results in a magnetic attraction. If one moves quickly enough, the laws of nature described by Newton and Maxwell would predict that they would eventually attract each other more strongly than they would repel, creating a paradox: the rods cannot repel from Hal's perspective but collide from Barry's perspective.

The Einsteinian model resolves this paradox in a somewhat surprising way. The charge density on the two rods plays just as strong a role in generating the electrical force as it does the generating the magnetic force. As a result, every situation changes from a repulsive force to an attractive force at the exact moment the moving observer reaches the speed of light, which is not possible. (If Barry has a flashlight in hand, he will never outrun the light that comes out of it, so he must always be travelling at less than the speed of light.) True, Barry will observe a stronger force than Hal observes, but he also measures less elapsed time than Hal, so the overall effect results in the same predicted motion of the two rods.

Chapter 3

Space and Time

3.1 Space and Time Before Relativity

Prior to Einstein's work, the importance of both space and time to physics was clearly recognized. They were, however, treated as completely distinct quantities. On one hand, they had the spatial dimensions. These came in three fundamental directions. The scientist had the ability to choose these directions to some degree. On the surface of the Earth, one could choose North as one direction, East as another, and Up as the third. If one chooses North then one does not choose South; mathematically, they describe the same line, since you can "undo" all travel in the direction North by travelling in the direction South. Directions are only "different" if they cannot be completely cancelled by motion in the previously defined directions. One could choose Northeast as a direction, since the "East" portion can't be cancelled by North, but choosing directions that aren't at right angles to each other usually makes things more complicated than they need to be when working on the math. The directions are connected through rotations: if you rotate through a right angle, the person who was facing North can now be facing East or Up. It is easy to see how they relate to one another, and that close relationship is one of the reasons we have the ability and inclination to define whatever directions are convenient for us. (For example, when working on a ramp, it is more convenient to pick directions parallel and perpendicular to the surface of the ramp than simply North, East and Up.)

On the other hand, there was (and is) time. No clear connection between space and time was apparent. Sure, they could plot graphs using time as one axis and space on another, but that was a technique used to see how systems evolve, and was not considered an indication of any deeper, fundamental connection between the quantities. Time was absolute and the same (invariant) for all observers, while distances were different for observers in motion (though not as different as they are under Einstein; the distance a car travelled was variable under Newtonian mechanics, but the length of the car was not, since the front and back could be measured "simultaneously.")

3.2   Space and Time After Relativity

With Einstein's relativity, we learn that time is not invariant, and that perceived time intervals depend greatly upon the motion of the observer. Moreover, the manner in which one measures distances depends on time measurements and vice versa. This implies a much stronger connection between the two quantities than was previously believed. This begs a number of questions: if time is that closely linked to space, can it be treated as another direction? If so, why does that direction behave so much differently than space? Why can we not navigate back and forth through time as easily as we do in space? Why can't we simply rotate from a spatial direction to a time direction as easily as we rotate from North to East? These questions opened some very productive floodgates. Physicists were forced to examine their beliefs about geometry, which led to some startling revelations when they explored the abstract discoveries made by mathematicians in the recent centuries.

Mathematicians can always be depended upon to indulge their curiosity, regardless of whether or not their discoveries apply to the "real world." One such avenue was the exploration of the square roots of negative numbers. For centuries, it was believed that negative numbers could have no square roots. In fact, they can, although it was difficult for even mathematicians to accept that fact when they were first proposed. This resistance is what led to the square roots of negative numbers being named "imaginary numbers:" even the mathematicians at the time had a hard time accepting them.[1] Despite their name, imaginary numbers are just as valid as any other numbers. Moreover, mathematicians had found that one can include directions describing imaginary numbers in geometry and arrive at precisely the type of system physicists needed. If time were measured along an imaginary direction, then it would be a direction tightly linked to the spatial directions, and yet would still be distinct and unique in precisely the manner required to satisfy intuition. No simple[2] rotation can point someone in an imaginary direction, yet the direction is an integral part of the geometry of reality. This distinction is enough to explain why we cannot control our motion along the time direction: it is a fundamentally different type of direction. Still, this would have been incredibly difficult to accept if not for the discovery of invariants.

3.3   Invariants

The realization that this mathematical framework described the time direction went hand in hand with another accidental discovery that had yet to be explained, related to the lengths of certain mathematical objects.

In Newtonian terms, moving observers disagreed on the positions of objects, but they always agreed on the lengths of those objects. Relativity threw that convenience out the window, but did so in a way that pulled time in as another direction

---

[1]Most high school students have heard of the real numbers. They were named to contrast the imaginary numbers in an act of professional mockery. "You can study those imaginary numbers if you'd like, but *I* work with real numbers.

[2]This word is surprisingly important, as we'll see in section 3.4.

to be considered in the geometry. Physicists began working with various ways to combine the space coordinates with the time coordinate to see if this idea of an "invariant" length could be preserved. They had discovered a means to make this happen that was remarkably counterintuitive.

If an object is turned to lie along one of the directions chosen for our coordinate system, then it is relatively easy to measure its length: find the coordinates of the two ends, and subtract the smaller number from the bigger number. Since the days of Pythagoras, there was a known method for measuring the length of an object that didn't lie along one of these directions: measure the coordinates of the extreme ends of the object along each of the chosen directions, square them, add the squares, and then take the square root of the answer. The calculation may be somewhat cumbersome without a calculator, but it is useful to know how to do this when one cannot simply turn the object or the measuring device to measure something in the simple, intuitive manner.

In trying to do this with relativity, physicists noticed something very strange. Adding in the square of the time coordinate didn't resolve the issue: both time and distance were smaller numbers in a moving reference frame. The strange and startling discovery was that these reductions could cancel each other out if subtracted! In other words, if one added the squares of the lengths of the object along the space dimensions, but then subtracted the square of the length along the time direction, the answer one got was the same for all observers. Quantities that are the same for all observers are known as invariant quantities.

This idea was shocking and counterintuitive, but completely consistent with treating time as an imaginary number. Recall from the previous section that imaginary numbers were discovered as the square roots of negative numbers: if one squares an imaginary number, the answer is negative. Adding a negative number is mathematically identical to subtracting a positive number. The implications of relativity get crazier and crazier, and yet somehow remain entirely consistent with each other and with experimental results. Slowly but surely, Einstein's seemingly insane notions were gaining more and more support.

3.4 Minkowski

Hermann Minkowski laid much of the foundation for this combination of space and time when he addressed the attendees of the 80th Assembly of German Natural Scientists and Physicians[3] on September 21, 1908.[4] His speech began as follows (after translation):

> The views of space and time which I wish to lay before you have sprung from the soil of experimental physics, and therein lies their strength. They are radical. Henceforth space by itself, and time by

---

[3]No, the term "physicist" was not in use yet. Should you travel back in time by more than a century and develop some sort of medical ailment, be sure you know the true expertise of any physician you encounter.

[4]Minkowski was 44 years old at the time of his address, and would die of appendicitis in January 1909.

itself, are doomed to fade away into mere shadows, and only a kind of union of the two will preserve an independent reality.

Minkowski's address then continued to lay the foundations for the work done earlier on in this lesson, though it should be noted that the usefulness of imaginary numbers had yet to be realized; Minkowski subtracted the square of time, not because he treated time as imaginary, but because it was already known that this was an invariant quantity. Still, he laid a geometric foundation that is incredibly useful today: the space-time continuum.

The difficulty in graphing problems in relativity is that time can be treated as an imaginary fourth direction. We can graph two dimensions easily, and can approximate graphing three dimensions by "projecting" them onto a two dimensional page. Graphing four dimensions cannot be done on two dimensional paper with pen or pencil.[5] Minkowski found a means to graph space and time together without losing the unique qualities of time. He realized that we could not, as is traditional, keep all of our axes at right angles. These so-called Minkowski diagrams formed a picture to go along with the concepts, and opened the gates to allow physicists to envision the situations in a useful way. They could graph situations as measured by two different observers on a single graph, making their work far more efficient. This also added another tool to the toolkit.

With space and time this closely connected, and with multiple observers graphed on the same diagram, an even more startling discovery was made: there is a rotation that allows one to rotate from a spatial direction into a time direction! This rotation manifests itself physically in an unusual fashion: accelerating to a new speed is mathematically identical to rotating with respect to the time access. The "rotation" exists, and we do have the ability to implement it, but the imaginary nature of the time direction means it "looks" completely different than a spatial rotation to us.[6] We still cannot travel backwards in time, but Minkowski's work laid foundations to explain that as well.

3.4.1   Causality

It is an axiom[7] of science that cause precedes effect. Relativity seems to support this concept.

Imagine two events in the space-time continuum. One happens in a given place and time, and the other happens in a different place at a different time. One can calculate the invariant distance between these two events, in the manner described above. These calculations involve both positive and negative quantities, and the results can be positive, negative or zero. We say that the events are time-like

---

[5]Approximations can be made by creating multiple three dimensional graphs, each at a different time, but this is cumbersome and makes the time evolution of the system more difficult to see from the graph alone.

[6]It has been proposed that there are life forms which perceive time no differently than they perceive space, though no concrete evidence exists. It has also been proposed that said life forms live inside an artificial wormhole and are worshipped by the inhabitants of a nearby world that look identical to humans aside from their noses, so the science involved is highly questionable.

[7]An axiom is an idea that is accepted as true, but which cannot be formally proven. They typically haven't been formally disproved, either, as that would result in the loss of the axiom.

separated in one case, space-like separated in another, and light-like separated in the third. Time-like separated events are those which can be connected by cause and effect, while space-like separated events cannot. Light-like separated events form the border between these two categories.

Let us examine explicit examples. Imagine one drops a rubber ball on the floor, and you consider the two events in question to be the first time it hits the floor, and the second time it hits the floor (after bouncing.) These events are time-like separated because they happen at the same point in space; it is only time that separates them. The cause-effect relationship is possible and established by observation. Any two time-like separated events find that the square of the distance between them in the time direction is greater than the sum of those in the spatial directions. As a result, information about one event can travel at the speed of light at reach the spatial coordinates of the second event before the second event takes place.

Now imagine two people are bouncing balls, one on planet Earth, and the other on the rock formerly considered a planet which is named Pluto. If an observer in a spacecraft half way between them believes they are bouncing their balls simultaneously, then this observer will say that there is a separation in space, but not in time. These events are space-like separated; neither person can have any effect on the other person's ball bouncing activity. There is a difference in space, but not in time. Information travelling at the speed of light can leave one event the instant it happens, but by the time it reaches the spatial coordinates of the second event, the event has already happened. There can be no cause-effect relationship.

In the case of light-like separated events, the cause-effect relationship is possible, but the calculation of the four dimensional distance between the two events works out to exactly zero.

These ideas were not terribly new. What was new is the mathematical implications that seemed obvious in one of Minkowski's graphs: events that are time-like separated from the perspective of one observer are time-like separated from the perspectives of *all* possible observers. Similarly, space-like separations are always space-like, and light-like separations are always light-like. The implication supports the idea that cause and effect relationships must be preserved in space and time, and that "time travel"[8] is impossible. This idea will be revisited when gravity comes into play in a later chapter.

---

[8]Here the term "time travel" is used to mean travelling from an arbitrary point in time to another, equally arbitrary point in time, with little or no perceived time lapse between them.

Chapter 4

Paradoxes of Relativity

4.1 Introducing the Paradoxes

When Einstein's theory of relativity was first published, its counter-intuitive implications were met with extreme levels of doubt and scepticism. Early researchers
actively sought out logical contradictions in the framework that could be used to
justify a wholesale rejection of the theory so that it might be replaced with something more aesthetically pleasing. Some of these "paradoxes" are described here.
Study of the paradoxes is remarkably illuminating, as it forces the student to directly
confront intuitions and prejudices developed from years of working with Newtonian
mechanics.

4.2 Length Contraction

The length contraction paradox is one of the earliest paradoxes. Imagine you have
a sports car that is 5 meters long, and your garage is only 4 meters deep. Well, if
"moving objects contract," as Einstein's theory predicts, then you should be able
to fit your sports car in your garage, provided you are moving quickly enough that
the moving vehicle contracts to a length that is less than 4 meters. Everything is
fine until you stop the vehicle.

The paradox comes from the equivalence of all reference frames. From the perspective of the driver, the car is stationary, and it is the garage that is in motion. In
that case, the 5 meter car is being approached headlong by a garage that is less
than 4 meters long: the garage is still smaller than the car. An observer standing
near (or in) the garage, at rest with respect to the garage, concludes that the car
fits in the garage; an observer in the car concludes that the car does not fit inside
the garage. Now, the car is either in the garage or it is not, in all reference frames.
So, which is it? This was one of the "thought experiments"[1] that seemed to cause

---

[1]A "thought experiment" is an experiment one thinks about but doesn't actually do. This causes
a dramatic improvement in lab budgets, allows a tremendous opportunity to explore implications
of a theory, and severely limits the ability to make quantified observations to verify a theory.

problems for Einstein's theory.

4.2.1   Relativity of Simultaneity

The key to resolving this paradox lies in something known as the relativity of simultaneity. Two events are considered simultaneous if they happen "at the same time." It is not immediately obvious to most people that this concept depends greatly upon how we measure time. If time is measured differently by two observers, will they agree on whether or not events are simultaneous?

Let us examine what we mean when we say two events happen "at the same time." This means that information about these two events *reach the observer* at the same time. To pick a more mundane example, picture two different patrons at an outdoor concert in a very large venue. The stage has a large screen at the rear with live video of the vocalist, and most of the speakers projecting sound to the audience are at the front. The patron near the stage believes that the sound and image are simultaneous, as the lips and lyrics are synchronized on screen. The patron at the rear of the venue believes that the sound is out of synchronization with the video because the light from the screen brings that patron information about the video a short time before the sound from the speakers reaches the patron. This is true in Newtonian mechanics as well, simply because light is faster than sound, so that information arrives faster. They do not appear to be simultaneous.[2]

In relativity, we can pretend all information travels at the speed of light. A somewhat analogous situation arises in this case. Events are simultaneous for an observer if the information about those two events arrives at the same time. Since all information travels at the speed of light, this amounts to saying they travelled the same distance to the observer. This is easy enough to imagine when nothing is in motion: the observer is exactly centred between them.[3] What if the observer is in motion? What if one or more events are in motion?

Imagine that the events take place at locations A and B. To an observer who is at rest relative to those locations and is exactly half way between them, the events are simultaneous. Now imagine an observer in motion from location A to location B. This moving observer will only consider the events to be simultaneous if she or he is the same distance (equidistant) from both locations when information about the two events arrives. So, if the moving observer is exactly on top of our stationary observer when the information arrives, both observers will agree that the events are simultaneous. However, if the moving observer is not in this symmetric position,

---

Thought experiments are best used to design experiments to be done physically later.

[2]This analogy, of course, derives from limitations of human perception. If we could accurately perceive time in smaller intervals, the patron closer to the stage would also recognize that the events are not simultaneous. The only observer who could say the two events are simultaneous would be the observer who eyes and ears are pressed up against the face of the vocalist. This is not a viewing location recommended to those who want to watch the entire concert without being dragged off by employees of the venue's security force.

[3]Technically, the observer could be anywhere on the infinitely large plane between the two events. Imagine that the two events that occur are centred along the outside edges of the pages of an open book. The observer could be anywhere alone the spine where the pages meet. This book can be as tall as you'd like, and can be spun around. The observer can be any place the spine can be. This forms the infinite plane.

then information about the two events will arrive at two different times, and the moving observer will not consider the events to be simultaneous.

4.2.2 Resolving the Length Contraction Paradox

Somewhat contrary to intuition, the definition of simultaneity plays a large role in the length contraction paradox. Upon closer inspection, the paradox arises because the mental images it conjures for the driver involve knowledge about both ends of the car simultaneously. In other words, the transmission of information is considered instantaneous instead of delayed, violating the postulates of relativity. The solution to the paradox comes through in a manner that surprised the opponents of relativity: when the car enters the garage, the front of the car strikes the rear garage wall. Information about this collision is then transmitted[4] to the rear of the car. In the time it takes this information to reach the rear of the car, the rear of the car has already entered the garage;[5] both observers agree that the car fits in the garage! Both observers also agree that the driver is insane and the car is a write off.

4.3 Twin Paradox

The equivalence of reference frames also leads to problems with time dilation.

Imagine you are a twin.[6] At present, you are both the same age.

The paradox begins when you load your twin into a rocket ship and launch him or her on a round trip to another solar system a few light years away. Your twin is in motion the entire trip, so as far as you are concerned, time moves slower in her or his frame of reference. When the craft returns, you will be the older twin. Now imagine your twin's perspective. Your twin can consider himself or herself to be at rest the entire trip, so you should be ageing more slowly, and your twin will be the older twin upon her or his return. Which of you is older?

4.3.1 Resolving the Twin Paradox

The key to this paradox is the recognition that we are not, in fact, dealing with inertial reference frames. Recall the definition from our previous lesson: an "inertial" reference frame is one in which the origin of the coordinate system is not being accelerated. Acceleration includes direction as well as speed. When the twin in the rocket turns around for the return trip, she or he changes from one inertial

---

[4]The mechanics of transmission are irrelevant to the paradox. In the physical world, the information is transmitted via vibrations in the bonds between atoms and molecules in the car, and is transmitted at less than the speed of light. A lower transmission speed assists in the resolution of the paradox, and will be ignored in the main argument.

[5]This is guaranteed. If you look carefully at the math, the condition that says the car is fast enough for length contraction to make the car fit in the garage from the perspective of the observer in the garage is identical to the mathematical condition that the rear of the car enter the garage before information about the collision arrives. The car fits or doesn't fit when measured by all observers.

[6]If you really are a twin, imagine this is one of the days that you and your twin are getting on each other's nerves.

reference frame to another. In reversing direction, the pretence of simultaneity is broken, and the world proceeds as usual.

Complete details are available in the version of this lesson that includes the math. The crux of the solution is this: when the twin is rocketing away from Earth[7] both are correct in treating themselves as the "older" twin. If each twin sends signals to the other at regular intervals (such as birthdays) then both twins send more signals than they receive while they are moving apart. When the twin in the rocket reverses course, however, the timing is disrupted. This twin begins to see signals more quickly as he or she approaches their point of origin. By the time both twins are back on their home world, the twin in the rocket has received more signals than she or he has sent, and they agree that the Earthbound twin is older.[8] This does leave one more question, though: what is the age of the twin who made the round trip? For that, we measure the proper time. Proper time is not "more correct" than any other time, despite the instinctive reaction to the name. It is the time as measure in the rest frame of the moving twin. This is the amount of time that lines up properly with that twin's natural ageing process.

These are not the only "paradoxes" of relativity that one can find. Thankfully, all are resolvable: every seeming paradox proposed to date has been found to be reconcilable, and most often "arise" by applying our normal, day-to-day intuitions which are actually in error, rather than the theory of relativity itself.

---

[7] Assuming our readers are on Earth. If you are reading this on another planet, please e-mail me at `w.blaine.dowler@gmail.com` and let me know!

[8] This is punishment for sending your twin on a long term interstellar trip, I suppose.

Chapter 5

Energy and Momentum

5.1 What Is Energy?

One of the quantities that physicists find most useful is energy. It is often difficult to define for new students, due in large part to the fact that it is an intangible quantity that you cannot see, touch, taste or smell. The formal definition is often "energy is the ability to do work," which seems clear and straightforward until one is asked to define "work" in the scientific sense.

An alternative definition of energy is "the ability to cause changes in motion or position." From a scientific perspective, this is equivalent to the first definition, but it reduces the terminology to that which is intuitive for new students. There are different types of energy in the world, though we will only deal with two of them in this text.

5.1.1 Kinetic Energy

The type of energy that is often the easiest for students to accept is kinetic energy, or energy of motion. If an object is in motion, its own position is constantly changing. It also has the potential to cause changes in motion to other bodies through collisions.

In the Newtonian view, kinetic energy can be calculated based solely on the mass and speed of an object. If two objects with different mass travel at the same speed, then the object with greater mass has greater kinetic energy. If two objects with the same mass travel at different speeds, then the one with the greater speed has the greater kinetic energy. Moreover, the speed of the object has significantly greater impact than the mass. (If you double an object's mass without changing its speed, you double its kinetic energy. If you double its speed, you more than double its kinetic energy.)

5.1.2 Potential Energy

The second major type of energy is potential energy. This is a little harder to see, as there is nothing visibly moving or altered in any way as a result of obtaining potential energy. This energy can be "released" and turned into kinetic energy in one form or another. For example, a book on a shelf has gravitational potential energy: the book might fall, gaining speed and kinetic energy as it approaches the floor. A book on a lower shelf has less potential energy, as it cannot fall as far and will be travelling at a lower speed when it hits the ground.

Potential energy is a consideration when there is some sort of outside field or force that can drive changes in motion, such as gravity. As such, further discussion will be saved for the general theory of relativity. The variables one needs to describe potential energy depend upon the source and nature of the outside force or field.

5.2 What Is Momentum?

Momentum is a term that is commonly used in day to day speech in a fashion that differs from its scientific meaning. When people talk about momentum in the colloquial sense, they typically mean something is in motion and hard to stop. This usage is closer to the word "inertia" than momentum. It is inertia that determines the resistance to accelerations and changes in motion. Momentum, in the Newtonian view, is the product of mass and velocity.[1] In Newton's day, it was referred to as the "quantity of motion." Momentum is always conserved in a collision, meaning that the total momentum that goes into a collision is the same as the total momentum that comes after. (Again, direction matters; two vehicles meeting head on can turn into a stationary wreck immediately after collision because momentum in one direction, say North, can be cancelled out by momentum in the opposite direction, South.)

5.3 How Are They Connected?

Kinetic energy and momentum can both be calculated using the same two pieces of information: the mass of the body in motion, and the velocity of the body in motion. Although they are not identical quantities,[2] the fact that they depend solely on the same two variables begs the question about whether or not the two are related.

5.3.1 The Newtonian View

In the Newtonian view, energy and momentum are connected through forces. When a force is applied to a object at rest, it accelerates to some velocity. The force applies over a period of time, during which the object travels some distance. Assuming the object is in space, so we don't have to worry about friction and other factors, allowing the entire force to apply to accelerating the object in the exact direction

---

[1] It uses velocity, and not speed, because direction matters.

[2] They cannot be the same: direction matters to momentum but not kinetic energy, which is a significant difference on the conceptual side.

of the force, then we can calculate both the energy and momentum using simple and similar math. We will also assume that the force is uniform, meaning it applies with the same strength and in the same direction the entire time it is applied. To calculate the energy of the object, one multiplies the force by the distance it has travelled. To calculate the momentum of the object, one multiplies the force by the time elapsed during the acceleration.[3]

5.3.2 The Relativistic View

In the relativistic view, there are a few changes that need to be made to these concepts. The first and foremost idea that needs to be explored is whether or not there is an upper limit to either quantity. After all, the universe has a speed limit in the speed of light. If these quantities depend on the speed of an object, then one would tend to think that there could be an upper limit to either or both in turn.

The relationship between the two quantities, as force applied along different directions (space or time), still applies in relativity. In that case, conceptually, we see that there should be no upper limit to either quantity: one can continually apply a force to an object, regardless of how much time elapses or distance is travelled. So, how do we obtain unlimited energy and momentum with a limited speed?

The answers lies in the fact that we have misidentified the variables involved. As we mentioned in lesson two, Newton originally recognized inertia as a key quantity, and then proposed that the mass and inertia of an object were one and the same quantity. Although he couldn't possibly have realized this at the time, mass and inertia are not, in fact, the same. The inertia of an object increases as its speed increases. This increase is almost imperceptible at the speeds that we see in our everyday experiences, but it is unmistakeable at higher speeds. As an object gets closer to the speed of light, its inertia increases incredibly rapidly, ensuring that no amount of energy or momentum will carry the body above the speed of light. In several of the cases in which we used mass in Newtonian mechanics, we should have used inertia as Newton originally proposed. This concept was so counter intuitive, and the notion of equality between mass and inertia so widely accepted, that when this difficulty was first noticed physicists incorrectly assumed that the mass of an object was increasing, and the quantities were named as such. The mass was referred to as *rest mass* and the inertia was referred to as *relativistic mass*. Although the error in these terms is now understood, they were in use for so long that they still appear in much of the literature on the subject.

The next question pertains primarily to momentum: with momentum, direction matters, so what quantity appears as the time component of momentum? We now have four directions that need to be described when dealing with quantities that use direction at all. To discover this component, we begin with Newton's definition of momentum as the product of inertia and velocity. We take our four dimensional velocity and multiply it by the inertia of an object. What we find in the time component of the result is startling: it is the kinetic energy of the object, with an

---

[3] If the force is not uniform, some adjustments need to be made. Instead of simple multiplication, we use calculus to integrate the force with respect to the other variable.

additional term added: mc^2, where m is the mass of the object and c is the speed of light.

This was utterly shocking. An object at rest contained energy in the absence of outside fields or forces, and that energy E had the quantity mc^2. Never before had it occurred to anyone that mass could somehow store energy, and yet it seemed apparent that mass was some form of condensed, solidified energy. This revelation is the one that led to nuclear power. It was so unexpected that Einstein himself discussed it tangentially in his paper, deviating from the calculation he worked on to make a special point of describing it. It is also one of the primary examples of serendipitous discovery in the sciences: who would have believed that nuclear energy would have been discovered by examining momentum in a universe with a speed limit?

Chapter 6

Electricity and Magnetism

6.1 Electricity and Magnetism the Newtonian Way

We have already seen how the laws of electricity and magnetism led to the rise of relativity. Let us examine how relativity alters them more closely.

Both electrical and magnetic phenomena have been studied for thousands of years. It was not until 1820, however, that the scientific community at large recognized how closely connected the two are. Hans Christian Oersted noticed (which giving a lecture to students) that electric currents deflect the needle of a compass. Further studies were able to quantify the relationship, explicitly defining and mathematically defining the procedure. It was clear that an electrically charged object in motion produces a magnetic field that is circular in shape and surrounds the particle at a right angle to the direction of motion. There was a problem, though: there was no *theoretical* reason to connect the two interactions. The two types of phenomena were kludged together mathematically because the connection was experimentally proven, and *not* because of any intrinsic motivation to tie the two together. It was comparable to rain and rainbows: people recognized that rainbows appeared after rainfalls so early on that even the word "rainbow" includes the word "rain," but it would be centuries before the optics involved were understood well enough to explain precisely why rainbows appeared.

The theoretical descriptions of electrical and magnetic phenomena was completed in the late 1860s, and James Clerk Maxwell collated the results into what are now known as Maxwell's equations. These described all known electrical and magnetic phenomena, detailing the connections between the fields without explaining why they are connected, with particular focus and attention on how the electric and magnetic fields change and evolve through space and time.[1] Just as ancient people could predict the appearance of rainbows, people in the 1900s could predict the magnetic fields produced by electrically charged objects to great effect, but the connection was arbitrary. This was still a young science, so when it was shown

---

[1]Remember, to James Clerk Maxwell and his contemporaries, space and time were two completely distinct quantities.

to be inconsistent with Newtonian physics, it was assumed that the fault lay with electricity and magnetism. As both theories were consistent with the experimental data available at the time, this assumption seemed reasonable. As we learned in earlier chapters, it was actually the Newtonian mechanics that were at fault.

6.2   Electricity and Magnetism the Relativistic Way

The work of Lorentz and Einstein combined to solve the problems with electricity and magnetism in multiple frames of reference. Furthermore, they showed that electromagnetic theory did not need to be altered in any special way; the framework established by Maxwell's equations was completely consistent with electricity and magnetism. There were, however, other significant benefits to the relativistic framework that were worth exploring.

So far, when we take a relativistic viewpoint of quantities, we have found a common pattern: quantities that depend on direction get a fourth quantity attached which serves as the "time direction" component of the quantity. In all such cases, the fourth quantity was independent of direction. In the case of electrical fields, the prime candidate for such a fourth quantity is the magnetic field, but that also depends on direction. It can't fit easily into that fourth slot. This seems to be a stumbling block on the path to connecting electricity and magnetism on the theoretical side.

Undeterred, we go back to what we learned in chapter four: although no simple rotation can move something from a spatial direction into the time direction, one can formulate a rotation based on acceleration that does the same job. This becomes the key to connecting the two phenomena.

Imagine, if you will, a long, electrically charge rod that is at rest. In the classical view, accelerating this rod along its axis[2] causes no change in the electrical field, but produces a sudden and inexplicable magnetic field. If we had two such rods lying parallel and moving in identical directions at identical speeds, the magnetic force created would be an attractive force. We now know that such a motion will increase the electrical force: as the rods accelerate, their lengths contract. With the same amount of charge packed into a shorter distance, we see the electrical charge density increase, which causes an increase in the electrostatic attraction. By carefully calculating the result of "rotating" the electric field to a higher velocity using the math that describes our relativistic world, we find that the action of the classical magnetic field is *identical* to the action of this "extra" electric field produced by relativity! This math is independent of the situation, so it doesn't matter if our moving electrical charge is a rod, disc, point, etc. This is true for any shape that moves while carrying an electrical charge.

This was something of a holy grail to physicists: the elusive theoretical connection between the electric and magnetic fields that didn't exist in the classical view became an immediate and natural consequence of relativistic dynamics. The magnetic fields observed around active circuits resulted directly from the motion of charged particles

---

[2]By "along its axis," we mean that if the rod lies in the North/South direction, then we accelerate it either North or South, and not East, West, up or down.

under the rules of relativity. The mathematical object used to combine the two is more complicated than a four directional vector, but it is just as valid: the fields were unified into a single phenomenon, and Maxwell's set of four[3] equations simplified to a single equation. This served as yet another triumph of the special theory of relativity. Despite a rocky start, the scientific community had finally accepted special relativity as the theory that explained reality better than any other theory. It is now known that any errors in relativity are small, as the theory has been tested with incredible experimental rigor. Still, special relativity is limited to points of view that do not experience forces or accelerations. Most of our work is done in the presence of a gravitational force, and many real world observers need to change their direction to do their jobs. The special theory of relativity was inadequate for these tasks, and would need to be expanded into the general theory of relativity, which is the subject of our final chapters.

---

[3] Well, technically, Maxwell did his work before physicists adopted vectors as mathematical tools, so his "four" equations were originally published as twelve equations. In a roundabout sort of way, this also explains why the symbols for so many electromagnetic variables are counter intuitive, such as using B for the magnetic field instead of M.

Chapter 7

The Postulate of General Relativity

7.1 Terminology: Special and General

What makes "special relativity" special? What makes "general relativity" general? These are common questions when one first researches the subjects, particularly when one is coming from a non-mathematical background.

In mathematical situations, research is often devoted to special and general cases. A "special" case is one in which specific values are known, or specific restraints are applied. For example, most elementary school math deals with special cases. Students solve questions such as 234×123 instead of solving the general case of $a \times b = c$ and then substituting $a = 234$ and $b = 123$ before solving for c.[1] Science works in a similar way. Special relativity considers the special case in which the reference frames one uses to measure from are experiencing uniform, or unaccelerated, motion. General relativity expands the theory to allow observers to move with arbitrary accelerations.

---

[1] It is the author's opinion that the lack of explicit instruction about the usefulness and need for general cases rather than exclusively special cases is one of the main reasons students struggle with algebra. When faced with a massive paradigm shift in mathematical thinking that is not treated as a massive paradigm shift for fear of scaring students away, the students become convinced that algebra is an exercise in futility and harder than it needs to be. When the author taught algebra to first time learners, he started with lessons about the paradigm shift and the axioms of algebra and the need for special cases before getting into the nitty gritty steps. He then gave out a homework assignment asking students to manipulate high school physics formulae in the general case only, and the class average was 87%. When he later taught high school physics to a completely different group of students and gave those students the identical assignment, the class average was 63%. End rant.

7.2   The Postulate of General Relativity

The distinction between special and general relativity boils down to a single realiza-tion: one cannot distinguish between different accelerations by feel alone. Imagine being placed in a sealed and soundproof room (with ample oxygen supply.) Imagine also that you can distinguish between "up" and "down" as directions. There is a particular side of this room that you can call a "floor." Einstein realized[2] that there is no physical sensation that can be used to distinguish between standing in a sealed room on the surface of a planet and standing in a sealed room with a rocket engine attached that forces the room to accelerate.[3] This equivalence led to the formulation of non-inertial reference frames, or accelerated reference frames.

7.3   Gravity

Beyond allowing the formulation of non-inertial reference frames, this postulate connected gravity to reference frames, which were already connected to the speed of light. This was a monumental connection. At the time, only three forces were recognized in science: there was the electrical force, the magnetic force and the gravitational force. With the first two bound by relativity, scientists began searching for a unified theory that could bind the "final" force to the other two.[4] Developing this "Grand Unified Theory," or GUT, became Einstein's unrealized goal in life. Over 50 years after his death, such a theory has still not been formulated to the satisfaction of most physicists.

This formulation did, however, spawn a new revelation: mass and energy do not merely exist "in" the geometry of reality, they alter and define the geometry of reality. Imagine you are still in your earliest school years, and have been tasked with drawing a triangle. The implications of Einstein's postulate are akin to watching your piece of paper change and transform in shape and size as you draw the triangle on its surface. A flat piece of paper ceases to be flat as soon as something is drawn on it, instead curving in a way that attracts all other shapes on the page to the one you have just drawn. It is this revelation that has led to the notion of the "fabric" of spacetime. The mental image conjured by the mathematics renders reality as a great sheet. Massive and/or energetic objects dent this sheet, altering the "fabric" of the sheet of reality.

7.4   The Geometry of Reality

This, once again, changed the way the geometry of the world was laid out. "Ro-tating" objects to different speeds caused ripples in the sheet, moving and shifting

---

[2]He was riding an elevator at the time.

[3]We are forced to assume that any noises or vibrations caused by the rocket engine are not transmitted to the room.

[4]By today's count, there are four forces: the gravitational force, the electromagnetic force, the weak nuclear force and the strong nuclear force. Because electricity and magnetism are already connected and unified in our day to day situations, they are now treated as a single force. They have been unified with the weak and strong nuclear forces as well, but this unification is only visible under the extreme conditions created within particle accelerators,

the dents in what we perceive as gravity. This even led to the theory of gravitational waves, which emanate from a source of gravity and propagate throughout the universe.

These dents and ripples have a great impact on the motion of objects. It is no longer natural to think of objects thinking in straight lines. They can move in the straightest possible lines, but that isn't quite the same thing; one can drive a "straight" stretch of road for several kilometers, but odds are the road curves along the surface of the Earth as elevations change on the oblate spheroid and so they are still treated as distinct forces in most circumstances. we call home. We now talk about "geodesics," which are the straightest possible lines one can follow on a curved surface.

To thoroughly drive the strangeness of this concept home, we will look at a shape that cannot exist on a flat surface, but which every reader is likely familiar with without even realizing that this is the case. The shape we are talking about is a "biangle." This shape is formed when two lines are drawn as straight as possible (i.e. two geodesics) and meet in exactly two places. In the geometry taught in public school, this cannot happen, but the geometry taught in public school is restricted to the "special case" of flat surfaces. The "general case" of geometry allows this shape to exist. You will probably not find examples in your mathematics classroom, but are extremely likely to find them in your geography classroom.

Take a globe of the Earth. It is covered with lines of longitude and latitude. Some, but not all, of these lines are geodesics. We need to determine which are which. Imagine you have a globe from the factory before it has been painted or marked in any way. It is a perfect sphere. You take a marker and start drawing the straightest possible line that you can from any point on the surface. You will eventually return to the spot you started from.[5] As the globe started with no distinguishing features, all such geodesic lines would be identical when drawn carefully and accurately. This is the case with lines of longitude on a globe, and not most[6] lines of latitude. If one takes two lines of longitude for examples on the globe, you will find that they form a biangle; they meet at the north pole, continue along the straightest possible paths until they are parallel while crossing the equator, and then converge to meet at the south pole and complete the biangle.

Similarly, "rules" for shapes that we are aware of do not necessarily apply. For example, we are taught that all triangles have angles which add up to 180°. While this is absolutely true on a flat surface, one can easily form a triangle with three right angles on an appropriately curved surface. Begin at any point on the equator. Travel directly to the north (or south) pole. Make a 90° turn and continue until you reach the equator once more. Make another 90° turn (in either direction, East or West) and return to your starting point. The triangle mapped out by these three geodesics will have three right angles, for a total angle of 270°.

We can take this same issue one step further and demonstrate the "parallel transport

---

[5]Note that the author assumes your penmanship and artistic skills are significantly better than his own. He'd miss the target by a significant amount.

[6]There is a grand total of one line of latitude that forms a geodesic. Deducing which line that is indicates a good understanding of the topic.

problem." Again, imagine you are standing on the equator, facing North with your right arm held out in the Eastward direction. Walk to the north pole, and with each step, you keep your arm held out parallel to the direction is was in when you took your previous step. Upon reaching the north pole, turn right, keeping your arm parallel to its original direction so that it now points directly ahead of you. Walk until you reach the equator and turn right again; you will now either have your right arm across the front of your body, or you will extend your left arm to maintain the same direction.[7] Walk back to your starting point. Your extended arm has been held parallel to its previous direction every step of the way, and yet the arm that started out pointing East is now pointing South! This is the parallel transport problem: you cannot transport something pointed a particular direction and guarantee that its direction will remain constant across a curved surface. Those paths which keep their own direction constant (i.e. those that don't "feel" like turning to a walking person: if you point your arm straight ahead with your first step, it points straight ahead of you at every step; if you return to your original position, you are facing in your original direction when you do) are the geodesics for that surface.

The concept of the geodesic is not just one that seems correct, but is one that turns out to be fundamental to the behaviour of objects in our universe. If we go back to Newton's original axioms, he stated "Unless an outside force is applied, a stationary object (object "at rest") will remain stationary and a moving object will continue moving at the same speed and in the same direction." In short, things move in straight lines. We now modify this to say that, in the absence of outside forces, objects move along geodesics.

This counter intuitive geometry had been explored by abstract mathematicians for centuries before its application to reality through relativity. With this toolkit available, physicists soon discovered theoretical objects that still hold imaginations enthralled to this day: black holes and worm holes.

---

[7]Your right arm is probably pretty sore at this point.

Chapter 8

Holes in Space

8.1 Black Holes

Scientists and science fiction writers have been fascinated by black holes since before they even had names. Some of these writers did insufficient research, and as a result, popular understanding of black holes includes a number of misconceptions. Before we begin to examine the reality and the fiction surrounding these objects, however, we must first establish some quantum mechanics.

8.1.1 Quantum Mechanics

All objects with mass are made up of a collection of subatomic particles, the most common of which are electrons, protons and neutrons. It is less common knowledge that protons and neutrons are, themselves, made up of subatomic particles known as quarks, and that the right conditions can force a proton and electron to combine to form a neutron. The universe is also filled with "virtual particles," which appear temporarily, generally in pairs, and allow objects to feel forces, among other things. As surprising as it may be, these incredibly tiny objects serve important roles in the study of black holes. We will deal with virtual particles in section 8.1.4. For now, we will only examine the effects of great pressure on atoms and nuclei.

Stars are the most massive common objects in our galaxies. They release energy due to natural fusion reactions. In short, they have so much mass that the atoms in their cores are crushed together into larger particles. The heat energy this releases helps the star maintain its size and structure. As the atoms within the star get larger and larger, it requires more and more energy to force them to combine into larger atoms through fusion. Stars eventually "run out of fuel," meaning that the atoms in their core are too large to be forced to fuse together by the star's gravity any longer. The star "dies" as the internal pressure suddenly drops, setting off a chain reaction that leads to a nova or supernova explosion as the internal pressure suddenly drops. What remains is a massive husk, rife with gravitational pressures, but lacking the internal fusion that provides a counter pressure to keep the star at its original size. If this husk is large enough, it forms a neutron star; all of its internal atoms are gravitationally crushed together until the core is a single giant nucleus made solely out of neutrons. In some cases, the husk has too much mass to even exist as a neutron star, and it continues to collapse. These objects become black holes.

8.1.2   What Black Holes Are and Aren't

A black hole is an object that has so much mass its internal gravity is too great for even a giant nucleus to hold itself together. All of the particles within it continue to collapse into a point, also known as a singularity. This singularity has mass, but no volume; it is just a spot in space. It is this lack of volume that makes it dangerous.

The force of gravity felt between two objects depends on three variables: the mass of one object, the mass of the other object, and the distance between them. The closer the two objects get, the stronger the force of gravity between them. This is why gravity is always strongest when the objects are in contact, and the distance between the centres of the two objects is as small as it can get. In the special case in which at least one object is a planet, the force of gravity is strongest when the second object is on the planet's surface. Imagine Earth is the first object and you are the second object. If you were to start digging into the Earth, travelling down into and through the molten mantle, you would notice that gravity steadily decreased as you got closer to the core.[1] This is because much of the Earth's mass is "above" you as you tunnel, and the gravitational force from that mass is pulling you upwards. Now imagine the Earth has collapsed into a black hole. It has no surface that you can reach, so getting closer and closer just means the force of gravity gets stronger and stronger.

The stronger the pull of gravity, the harder it is to escape. The faster you are travelling, the better your chance at escape. (Rockets can escape Earth's gravity to visit the moon and other places, but people can't escape Earth's gravity by running up a hill and jumping as high as they can.) The key to leaving a planet is reaching "escape velocity," the speed at which one must travel away from the planet's core in order to escape its gravity. A black hole is an object with no volume, which means there is no limit to how close one can get to it. As a result, one can get close enough to it that the escape velocity at that point is greater than the speed of light. The point at which escape velocity equals the speed of light is known as the "event horizon."[2] As we know, the speed of light is woven into the fabric of our spacetime. The geometry of black holes gives them their name: if the "dent" made by the gravity of an object is so steep that not even light can escape, then for all intents and purposes, it punches a "hole" in the fabric of reality. In actuality, the fabric of spacetime still exists within the black hole, but it might as well vanish as far as those outside the black hole are concerned, as the interior of the black hole can never be observed.

---

[1]This is assuming you notice anything other than the intolerable heat.

[2]The etymology of this name goes back to the mind set from earlier chapters: relativity doesn't focus on objects at locations in space, but at events that occur at locations in spacetime. Just as one cannot see objects past the horizon on Earth, one cannot observe events that take place behind the event horizon of a black hole.

It is a common belief that, should our Sun be replaced by a black hole with mass equal to our Sun, our entire solar system would be swiftly sucked into this black hole and crushed beyond recognition. This could not be further from the truth. The key to the unusual effects of a black hole is proximity; if you aren't close to the event horizon, you don't notice a difference. If our Sun were to be replaced with such a black hole in some instantaneous manner, Earth's orbit would be completely undisturbed. The lack of incoming energy from fusion would still doom the human race when photosynthesis stopped and temperatures dropped, but it would be a slow and agonizing planetary death instead of the swift and exciting death predicted by underresearched stories. When you do approach the event horizon closely enough, you witness a number of unusual phenomena.

8.1.3   Black Holes Have No Hair

There are a number of quantities that can be tracked and conserved when studying most objects and events. For example, if car A contains a certain number of electrons, and car B has a different number of electrons, the total number of electrons remains the same after the two cars collide in the street. One can count electrons after the fact and figure out how many electrons each vehicle started with.[3] With a black hole, that and other information is lost. In fact, when studying a black hole, only four pieces of information about the matter that forms the black hole are preserved: the total mass-energy of the objects and particles involved, the net electrical charge of those objects, the net magnetic charge[4] of the objects, and the net angular momentum[5] of those objects. It is said that "black holes have no hair," using the metaphor that the lost information is equivalent to hair. It's a strained metaphor.[6]

8.1.4   Hawking Radiation

One of the reasons Stephen Hawking is so highly regarded amongst the physics community is that he was one of the first to examine the combined implications of quantum mechanics and general relativity, and in doing so he made an astonishing discovery: black holes evaporate.

It is a well known fact of quantum mechanics that "virtual particles" appear and disappear frequently. Hawking was the first to examine what happens when these particles appear and disappear near a black hole. It is possible that one particle

---

[3]I don't claim it's easy to count them, only that it is conceptually possible.

[4]Although theoretically possible at quantum mechanical levels, magnetic monopoles (north poles without south poles or vice versa) have never been observed. In theory, if they do exist and fall into black holes, these quantities will survive.

[5]Angular momentum is a combination of an object's geometry and rotation speed. A black hole retains this combination, but isolated information about either the geometry or rotation speeds of the objects the black hole is made of is lost.

[6]If the metaphor is so strained, why is it in use? Purportedly, the scientists who proposed the vocabulary did so because several non-English languages translate the term "black hole" into a term that is also used as a slang term that refers to a part of the female anatomy. They thought it was amusing to force that phrasing upon scientists working in those languages. This kind of attitude may be one of the reasons the field of physics is still so male-dominated today. The gender gap is smaller than it used to be, but it is still far too wide.

falls into the black hole, while the other escapes. These "virtual particles" typically annihilate with each other and disappear to conserve the total energy available. When one is captured by a black hole, it gives the black hole a *negative* amount of energy. Thus, the other particle can continue to exist as a real particle, and the black hole *loses* energy. The smaller the black hole, the faster this takes place. This is what has some people in a panic about particle accelerators: they fear that the creation of tiny black holes will somehow threaten the planet, either because of the energy released by evaporating black holes, or because the tiny black holes themselves are going to somehow consume the planet.

We have already discussed the latter fear; black holes don't suddenly suck everything up. This is a very good thing, as all elementary particles (electrons, the quarks protons and neutrons are made of, etc.) have zero volume, meaning every subatomic particle in existence *is* a black hole. Not big ones, but black holes nonetheless. As for the former fear, energy is still conserved, so a black hole that is made out of six high energy particles can only release the energy contained in those six particles while evaporating. Thus, it can only release energy in quantities less than or equal to the energy that was already contained by the particle accelerator and detector before the black hole was created.

8.1.5   Frame Dragging, Event Horizons and Ergoregions

Mass and energy have been tied tightly to the very fabric of spacetime. We know that a moving object creates ripples of gravity in this spacetime. What happens if it rotates rapidly, creating a lot of motion-ripples (gravitational radiation) in a single location?

The answer is a phenomenon known as "frame dragging," and it is not limited to black holes. When a massive object rotates rapidly, it can pull spacetime along with it. If you are a stationary observer positioned above this source of gravity (likely a star) you would not fall straight toward the object. Instead, you would be "dragged" around the object in the same direction it is rotating as you fall, which is why the phenomenon is called frame dragging. This is the phenomenon that explains the unusual nature of Mercury's orbit, first mentioned way back on the first page of our first lesson. Mercury's orbit rotates around the Sun because the spacetime around the Sun rotates *with* the Sun, dragging our solar system's closest planet right along with it. This effect is most pronounced near the object that is rotating.

Perhaps the most bizarre phenomenon surrounding black holes takes place inside the event horizon. When an observer falls inside the black hole, spacetime gets so twisted that the four directions of space and time get twisted around. Once past the event horizon, an observer is just as free to move through the time dimension as we typically can move through space. The trade-off is that the "down" direction takes on the position of the imaginary axis, meaning the observer is completely unable to control his or her inevitable descent towards the singularity at the centre of the black hole.

Because the effect of frame dragging is more pronounced as you get closer to the object, it becomes a significant effect when a black hole rotates. In fact, many black holes can rotate so quickly that they form "ergoregions" around them. In these regions, spacetime is being pulled so strongly that it becomes impossible to rotate around the black hole in the direction opposite to its rotation. If the black hole rotates clockwise (from your perspective) then it is impossible to form a close counterclockwise orbit around that black hole. Furthermore, the transformation and replacement of the "down" and "time" directions takes place within the ergoregion instead of behind the event horizon. When one falls through the ergoregion of one of these rapidly rotating black holes, and crosses the event horizon within, the transformation is reversed: within this black hole, time and space resume their normal roles! In fact, it is possible that the entire observable universe is contained within a supermassive, rapidly rotating black hole.

8.2 Wormholes

With the discovery that mass can warp spacetime, punching holes in it, scientists and science fiction writers got excited again. Perhaps it was possible to circumvent the light speed limit in some way: instead of propelling an object faster than light, one could bend and twist space until the destination is much closer, and then just move there at speeds less than light.

The first scientific suggestion related to this idea is commonly referred to as the "wormhole." Formally named the Einstein-Rosen bridge (as Albert Einstein and Nathan Rosen first proposed it in 1935 as a way to bridge two points in spacetime) the idea is that a tunnel, or hole, is created in spacetime by two black holes. Each black hole pulls the fabric of spacetime out of shape into an escapable region. It was proposed, however, that two such black holes, properly aligned, could create a tunnel that connected two far distant points in spacetime. Unfortunately, 27 years later, John Wheeler and Robert Fuller showed that such a body is unstable, and would collapse as soon as any mass or energy entered, turning the wormhole back into a pair of black holes before it could come to any practical use.

In 1988, Kip Thorne and Mike Morris showed that stable wormholes are only possible if some sort of exotic matter is held in place at the opening. While this does give some hope for faster than light travel, it is not much hope. The type of matter proposed has properties that have never been observed in our universe, and there have been no effective methods proposed for keeping that matter in place. Try as we might, it appears that the universe simply does not allow time travel to happen in the science fiction sense.

Chapter 9

Cosmology

9.1 What Is Cosmology?

Cosmology is the study of the structure and nature of the universe. Instead of focusing on any single object, cosmologists study the Universe as a whole and the way it evolves over time.

9.2 Einstein's Cosmological Constant

With the discovery of the general theory of relativity, the geometry of the universe was quantifiable and definable. This meant that, for the first time, the structure of the Universe could be studied explicitly. Einstein was amongst the first to do this calculation, and he ran into a result that disturbed him greatly. When he examined the geometry of the Universe we live in, he uncovered math that made it appear as though the Universe was expanding. This did not fit his ideas of theology: Einstein felt that the God he believed in would have created a perfect Universe to begin with, and expansion represented an undesirable form of change. The equations included a constant whose value could not have been measured at the time. Einstein arbitrarily chose a negative value for this constant which would result in a static Universe which would prevent this predicted expansion.

A few short years later, Edwin Hubble published his results which proved that the Universe was, in fact, expanding. This shocked Einstein and others. The Universe was an evolving beast of some form. Einstein and others recognized his earlier error and made another one: they arbitrarily decided that the cosmological constant should have a value of zero. This led to an expanding Universe model, and everything seemed fine. It seemed clear that, at some point in the past, the Universe had been a finite point, which then exploded in a Big Bang, and began expanding from that point.

This was the accepted model for almost 80 years. It was challenged in 1998, when a group of scientists recognized that the value of zero is just as arbitrary and any other value for the cosmological constant, so they set about trying to measure the value explicitly. What they discovered shocked just about everybody: the cosmological constant is not zero. Not only that, it is a positive number. In other words, the Universe would expand whether it contained matter or not. The question is: why? What physical process or object does this cosmological constant represent?

This is still one of the unanswered questions of cosmology. There have been a number of proposals, most of which involve dark matter or dark energy. Astronomical observations show expansion behaviour that cannot be accounted for by the matter and energy we can observe. Therefore, it has been proposed that there is some sort of matter or energy out there which we cannot observe, which is commonly referred to as dark matter and/or dark energy. Very little is known about dark matter, although much has been suggested. The difficulty with studying dark matter is that it is dark; it cannot be directly observed or studied using today's technology. This may change in the future as new technologies are developed, but that day still appears to be a long way off.

9.2.1   Expanding Into What?

One of the philosophical questions that has been asked is what the Universe is expanding into. Is there a larger universe with greater dimensions outside of this one? If so, what is that Universe "in?" Is there an infinite chain of universes, like some infinite set of matryoshka dolls?

The question about what is outside our Universe, if anything, will probably never be answered. Our instruments are limited to measuring that which is in our Universe, and that alone. It has been shown, however, that it is possible for the complete "exterior" geometry of such a Universe to remain unchanged, while the "internal" geometry continues to expand. If there is an outside Universe, our presence in that Universe may not appear to be changing at all.

9.3   The Ends of Time

One of the most commonly asked questions in cosmology is "how will the Universe end?" There have been a number of proposals for this.

- **The Big Crunch** - In this theory, the expansion of the Universe eventually slows, and the Universe starts to fall back in upon itself, destroying all life in the process.[1] This theory has a companion theory, in which the lifetime of our Universe is cyclic, and the Big Crunch would be followed by a new Big Bang, with a new Universe being formed. This would happen over and over again in an infinite cycle.

- **The Static Universe** - This theory proposes that there is just enough mass around to slow the expansion of the Universe without ever quite halting it. This, ultimately, leads to a Universe very much like the static Universe Einstein's theological instincts wanted. It wouldn't technically be static, but the

---

[1]Unless you are a being named Galan who likes to snack on planets. Then you might survive.

expansion would eventually become imperceptibly slow. In this case, thermodynamics predict that all energy would eventually be converted into unrecoverable waste heat due to entropy, wiping out all life in the "heat death" that results from all of existence boiling away into a plasma that no longer allows complex molecular structures to exist.

- The Constantly Expanding Universe - In this theory, the Universe simply continues to expand, causing an increasing space between civilizations and worlds, and reducing the average energy in any given volume. Life, in this model, would eventually die off when all temperatures in the Universe drop below the point at which chemical reactions can occur and all of existence is frozen out at temperatures very near absolute zero.

The discovery of a positive cosmological constant make the third theory the most likely theory. Tell your descendants to bundle up: it's going to get cold.

Bibliography

[1] Albert Einstein. *Relativity*. Routledge, 2013.

[2] Isaac Newton. *The Principia: mathematical principles of natural philosophy*. Univ of California Press, 1999.

[3] Bernard Schutz. *A first course in general relativity*. Cambridge university press, 2022.